KONSTANTIN GRCIC

Imagery by
Florian Böhm

NEW NORMALS

Verlag der Buchhandlung
Walther und Franz König

HAUS AM WALDSEE

In unserem Alltag finden sich häufig Objekte und Rituale, die ganz selbstverständlich wirken, aber noch vor kurzer Zeit kaum lesbar gewesen wären: Ein Smartphone auf einem Schreibtisch erscheint uns heute kaum bemerkenswert, für Betrachter*innen aus der näheren Vergangenheit stellt dieser flache Gegenstand jedoch ein unerklärbares Rätsel dar. Denn unsere jeweils neuen Realitäten werden immer wieder auch von Objekten und den daran geknüpften Verhaltensweisen verändert, und unser Leben wandelt sich durch ihre Nutzung und ihr Alltäglichwerden ständig. Zukunft und Gestaltung sind permanent miteinander verbunden.

In der Ausstellung Konstantin Grcic – New Normals im Haus am Waldsee werden wir mit dieser Idee des Alltäglichwerdens – eben mit kommenden Normalitäten – konfrontiert: Nicht die Gegenstände selbst erscheinen ungewohnt, sondern ihr Arrangement, die Kombination und ihr Verhältnis zueinander lassen uns nach zukünftigen Nutzungen und Ritualen suchen: Beim Gang durch das Haus begegnen wir unter anderem Konstantin Grcics Bell Chair (2020), der mit einer Fahrradkette an einen Rammschutzbügel aus Stahl gekettet ist. Der Tisch Champions (2011) besitzt nun integrierte Rückspiegel. Die Konstruktion Head in the Sky (2017), ein Ort, der die Möglichkeit bietet, im Freien zu arbeiten und zu denken, ist um Satellitenschüsseln ergänzt worden. Der Bürostuhl Allstar (2014) findet sich zwischen Baustützen an Gymnastikbändern befestigt. Im Obergeschoss des Hauses sind Grcics Bürostühle 360 Chair & Stool (2009) zu einem Karussell vereint, die goldene Traffic Chaise Longue (2013) ist gespickt mit Selfie-Sticks, und aus der Kombination der beiden Sofas Brut (2016), einem Gymnastikball sowie einem Netzgitter entsteht eine Art abstrakter Trainingsraum. In jedem Zimmer des Hauses hängt außerdem die Lampe Mayday (1999), die entgegen ihrer vielfach und mobil einsetzbaren Verwendung hier eine feste Halterung bekommen hat und nun wie ein gesetzlich vorgeschriebenes Sicherheitsutensil die Installationen und Environments der einzelnen Räume ergänzt.

Durch diese Rekombinationen von bekannten Objekten erleben wir in der Ausstellung Andeutungen, Skizzen, prototypische Anordnungen von New Normals – Vorgriffe auf eine der vielen Zukünfte, die uns erwarten könnten. Konstantin Grcic widmet sich in seiner Praxis als Designer stets diesen zukunftsgerichteten Ausgestaltungen unserer Lebenswelt und antizipiert in und mit seinen Möbeln und Objekten mögliche neue Gegenwartszustände. Damit verrückt er aktiv das Denkbare und überführt es in eine gestaltbare Sphäre. Grcics Stühle, Sessel, Tische, Leuchten, Regale und andere Objekte demonstrieren eindrucksvoll: Zukunft kann nicht gewusst werden – wohl aber gestaltet. Seine Entwürfe verschieben unsere Perspektive damit vom Bestehenden auf das Kommende.

In New Normals entfernen sich seine Designs nicht nur von ihrer eigentlich so offensichtlichen Funktion, sondern weisen eben auch auf andere, mögliche, neue Welten, in der diese Objekte ihr Potenzial entfalten könnten. Dennoch sind diese Versuchsanordnungen nicht als Prophezeiungen zu verstehen, vielmehr schärfen sie den Blick der Betrachter*innen auf unsere Gegenwart und die Rolle, die die Gestaltung unseres Alltags darin spielt. Welche Bezüge zum eigenen Leben, zu den Erwartungen an die Zukunft – dystopisch und utopisch zugleich – können aus diesen Szenarien gelesen werden? Welche neuen Betrachtungsweisen auf die Gegenwart ergeben sich aus diesen neuen Konstellationen?

Ludwig Engel und Anna Himmelsbach

KONSTANTIN GRCIC IM GESPRÄCH
MIT LUDWIG ENGEL UND ANNA HIMMELSBACH

WARUM HEISST DIE AUSSTELLUNG NEW NORMALS?

Konstantin Grcic: Der Titel bezieht sich auf den Begriff „New Normal", der mit der Pandemie ganz selbstverständlich in unseren Sprachgebrauch übergegangen ist. Als „New Normal" wird eine Form von Zukunft beschrieben, die eigentlich schon zur Alltäglichkeit geworden ist. Gemeint sind bestimmte Umstände und auch Dinge, die anzeigen, dass sich unsere unmittelbare Lebenswirklichkeit bereits verändert hat. Das Tragen einer Maske im Alltag stellt keine Ausnahme mehr dar, sondern ist zur neuen Normalität geworden. New Normals bezieht sich nicht direkt auf die aktuell vorherrschende Pandemie, sondern fragt über die Gegenwart hinausgedacht: Was könnte in Zukunft normal sein, und wie würde sich das auf die physische Welt der Dinge, mit denen wir leben, auswirken? Die ausgestellten Objekte und Installationen sind spekulativ. Es geht nicht darum, eine Vision für eine bestimmte Zukunft zu formulieren, sondern zukunftsgerichtete Fragmente möglicher kommender Normalitäten anzuskizzieren und auszutesten.

WIRD UNSERE ZUKUNFT DENN WIRKLICH SO AUSSEHEN, WIE DU SIE MIT DEN NEW NORMALS SKIZZIERST?

KG: Es war mir von Anfang an wichtig, die Ausstellung als Ausblick in eine der vielen möglichen Zukünfte zu verstehen. Design ist immer eine Projektion nach vorn. Meine ersten Recherchen orientierten sich an Szenarien, die von der Wissenschaft belegbar prognostiziert werden: der Klimawandel und seine Folgen. Aber, weil diese Thematik sehr politisch aufgeladen ist, hätte die Ausstellung eine Ausrichtung bekommen, die ich nicht wollte. New Normals ist eine Designausstellung. Was ich zeige und in welcher Form, bleibt frei. Die Exponate der Ausstellung bieten unzählige Anknüpfungspunkte dafür, sich selbst auf die Spurensuche nach neuen Normalitäten zu begeben. Die teils zweckentfremdeten Möbel, aber auch alle anderen in den Installationen verbauten Objekte und Materialien mögen beim ersten Betrachten fremd wirken, aber plötzlich könnten sie überall sein, und dann nehmen wir sie als normal zur Kenntnis.

ALS AUSGANGSPUNKTE FÜR ALL DEINE SPEKULATIVEN INSTALLATIONEN DIENTEN VON DIR GESTALTETE GEGENSTÄNDE. WAS HAT DAS MIT DEINEM VERHÄLTNIS ZU DIESEN OBJEKTEN GEMACHT?

KG: Als Industriedesigner entwickele ich Produkte von der Idee bis zur Serienreife. Dieser Prozess dauert in der Regel zwei bis vier Jahre pro Projekt. In dieser Zeit entsteht eine sehr enge Beziehung zu den Dingen – man könnte fast sagen, dass die Projekte wie Kinder sind, die ich großziehe. Sobald die Entwicklung abgeschlossen ist und die Produkte auf den Markt kommen, löst sich meine persönliche Bindung zu ihnen auf. Die Dinge leben ihr eigenes Leben, und ich wende mich neuen Projekten zu. So ist das, und so ist es okay. Manchmal wünschte ich mir dennoch, dass ich nochmals eingreifen könnte, um mit den fertigen Produkten zu spielen und sie weiterzudenken. Die Ausstellung bot mir die Möglichkeit, genau das zu tun. Ich habe meine eigenen Produkte verändert oder ihnen durch die Form der Inszenierung eine neue Deutung gegeben. Manche Produkte habe ich regelrecht umgebaut, andere habe ich durch fremde Elemente ergänzt. Im Computerjargon würde man diese Form des Eingreifens Hacking nennen.

GESTALTUNG ANTIZIPIERT OFT ZUKUNFT, INDEM ES DEM KOMMENDEN DANN SCHON FORM GIBT, WENN ES NOCH KEINEN NAMEN ODER NUTZEN HAT. OFT ENTSTEHT DAS NEUE JA GENAU IN SOLCHEN ALLTAGSMOMENTEN, IN DENEN DURCH UMNUTZUNG ODER VERÄNDERUNG DES BESTEHENDEN ETWAS BISHER NOCH NICHT SICHTBARES GESTALT ANNIMMT. SO FUNKTIONIERT AUCH GUTE SCIENCE-FICTION …

KG: Unser Bild der Zukunft entspricht schon längst nicht mehr der hyperdesignten Ästhetik von Filmen wie Star Trek. Blade Runner wäre eine passendere Referenz. Hier wurde die Zukunft als reale Welt dargestellt, an die immer wieder angebaut wird und die nur mit Mühe und Not und mit Duct Tape zusammengehalten wird. Dieses Bild der Zukunft ist unserer Welt gar nicht so unähnlich. Sie wird nicht auf einmal neu gestaltet, sondern verändert sich durch ständige Nachbesserungen und Eingriffe in den Bestand. Es gibt nicht eine übergeordnete Gestaltungsidee, es sind viele Dynamiken, die zusammenwirken und sich zu einem heterogenen Ganzen vermengen. Das meiste um uns herum wird auch in Jahrzehnten noch so sein wie heute. Unsere Städte und Gebäude werden nicht wie Raumschiffe aussehen, und auch das Haus am Waldsee wird einem ziemlich genau so erscheinen, wie es heute ist. Was die Zukunft von dem jetzt unterscheidet, sind viele subtile Details und kleine Veränderungen, die zusammengenommen unser Leben in den nächsten fünf, zehn und zwanzig Jahren anders prägen werden. Heute noch unvorstellbar, werden sie uns dann als ganz selbstverständlich erscheinen.

WENIGER ZUKUNFT, MEHR GEGENWART?

KG: Zukunft, die vorgibt, eindeutig zu sein, ist Kulisse. Die echte Zukunft ist unvollendet und bleibt in gewisser Weise immer etwas unheimlich. Das sorgfältige Beobachten alltäglicher Situationen dient mir als größte Inspiration für die Arbeit. Mich fasziniert die Wirklichkeit unserer Welt, wie sie ist. Diese zwingt uns Bedingungen auf, die ich als Designer kritisch hinterfrage.

WELCHE R■LLE SPIELT DIE DIGITALE WELT FÜR DICH IN DIESEM ZUSAMMENHANG?

KG: Im Computer ist alles möglich, es gibt keine Schwerkraft und keine Hürden der Machbarkeit – auch keine Verantwortung. Man kann etwas Großes, Schweres auf ganz dünne Beine stellen, es würde trotzdem halten.
Das digitale Austesten dieser Freiheiten ist inzwischen ein ganz wichtiger Teil meiner Entwurfspraxis geworden. Gerade weil es im Industriedesign immer um das Machbare geht, bietet mir

der Computer Möglichkeiten des Experimentierens, um über die Gewissheiten der Gegenwart hinauszudenken. Die Arbeit an New Normals war ein solches Experimentierfeld in physischer Form. Wir konnten Ideen in echt umsetzen, die normalerweise nur auf dem Bildschirm existieren würden. Für mich selbst wird die Ausstellung wichtige neue Zugriffe und Erfahrungen liefern, die ich in zukünftigen Projekten anwenden kann.

DU KURATIERST UND GESTALTEST IMMER WIEDER AUSSTELLUNGEN, IN DENEN DU ETWAS FREIER AGIEREN KANNST, ALS ES IN EINEM PROJEKT FÜR EINEN HERSTELLER MÖGLICH WÄRE.

KG: Ausstellungsprojekte sind nicht unbedingt freier, aber sie sind vielschichtiger als das Produktdesign. Was ich damit meine sind die unterschiedlichen Gestaltungsebenen, über die eine Ausstellung funktioniert. Vom inhaltlichen Konzept über die Architektur und Präsentation der Exponate bis hin zu Text und Medien. Als Ausstellungsgestalter fühle ich mich wie ein Regisseur, der das gleichzeitige Zusammenspiel vieler verschiedener Aktionen im Blick hat. Eine Ausstellung über die eigene Arbeit bringt noch eine zusätzliche Herausforderung mit sich: die Auseinandersetzung mit sich selbst. Für New Normals habe ich bestehende Arbeiten genommen und sie in etwas Neues umgewandelt. Die Exponate und Installationen, die in der Ausstellung zu sehen sind, erlauben es mir, meine Arbeiten noch mehr dahin zu verschieben, wo ich sie während des Entwicklungsprozesses gerne gesehen hätte. Und ich teste Szenarien aus, die während der Produktionsprozesse nicht stattgefunden haben.

SEIT WANN GEHÖREN AUSSTELLUNGEN ZU DEINER PRAXIS?

KG: Die ersten Ausstellungen habe ich Anfang der 1990er-Jahre im Auftrag der Firma Authentics, einem Hersteller von Haushaltsprodukten aus Kunststoff, gemacht. Es ging dabei um kommerziell motivierte Präsentationen im Rahmen von Messen. Meine Inszenierungen versuchten, die Produkte nicht als Verkaufsware darzustellen, sondern sie in ihrem kulturellen Kontext zu zeigen und ihre Geschichten zu erzählen. Wir haben damals mit minimalen Budgets gearbeitet. Die Projekte mussten sehr einfach und schnell umgesetzt werden. Aus heutiger Sicht waren das wertvolle Lehrjahre für mich. Ausstellungsgestaltung funktioniert wie ein gutes Bühnenbild. Die Gestaltung ist auf wenige präzise Elemente reduziert und vermittelt sich zu einem großen Anteil auch darüber, was nicht zu sehen ist.

HAT DAS HAUS AM WALDSEE ALS EINSTIGE PRIVATVILLA MIT SEINER SPEZIFISCHEN ARCHITEKTUR EINFLUSS AUF DIE AKTUELLE AUSSTELLUNG GEHABT?

KG: Ohne das Gebäude würde es New Normals in dieser Form nicht geben. Das Haus am Waldsee wurde in den 1920er-Jahren als luxuriöse Privatvilla konzipiert, was auch heute noch sichtbar ist. Da meine Arbeit wesentlich aus Möbeln besteht, war es eine ziemliche Herausforderung, an diesem spezifischen Ort eine Ausstellung zu machen, die nicht zur Einrichtung werden würde. Ich wollte das auf jeden Fall verhindern und trotzdem ein Bewusstsein dafür erzeugen, dass man sich in einem Wohnhaus befindet. Das Haus ist das Haus, und die Dinge sind die Dinge, und sie verhalten sich zueinander, ohne dass man auf die Idee käme, hier sollten Räume eingerichtet oder dekoriert werden.

WAS WIRD BLEIBEN VON NEW NORMALS?

KG: Ich hoffe, dass die Ausstellung starke Bilder produziert, denn die werden bleiben. Es sind die Bilder einzelner Exponate, aber auch deren räumlicher Inszenierung im Haus am Waldsee. Ich wünsche mir, dass sich die einzelnen Szenen zu einem großen Stimmungsbild zusammensetzen, das über den Ausstellungsbesuch hinaus nachwirkt. Vielleicht gelingt es, dass viele Besucher*innen meine Setzungen als Anstoß und Ermutigung verstehen, die eigenen Bilder über die Zukunft zu überprüfen und weiterzudenken.

Everyday life often throws up objects and rituals, which now appear perfectly self-evident, but would have been completely incomprehensible only a short while ago: a smartphone on a desk seems utterly unremarkable today, yet just a few years previously it would have presented an inexplicable mystery to any viewer. For our new realities are constantly changed by objects and their attendant behavioural patterns. Our lives are transformed by their use and us getting used to them. Future and design are inextricably, inherently linked.

The exhibition "New Normals" at Haus am Waldsee confronts us with this idea of objects becoming part of the fabric of everyday life – the idea of future normalities: it is not the objects that are unfamiliar. Rather, it is their composition, their combination and their relations among each other, which make us search for future uses and rituals. Walking through Haus am Waldsee, we encounter, among other things, Konstantin Grcic's Bell Chair (2020), which is fettered to a skirting protection by a bicycle chain. The table Champions (2011) now sports integrated rear-view mirrors. The construction Head in the Sky (2017), a place that affords the possibility of working and thinking outdoors, has been supplemented by satellite dishes. The office chair Allstar (2014) finds itself fixed between props by resistance bands. On the upper floor of the house Grcic's office chairs 360 Chair & Stool (2009) have been composed into a merry-go-round. The golden Traffic Chaise Longue (2013) bristles with selfie sticks. And the combination of the two sofas Brut (2016), a gymnastics ball and a mesh generates a kind of abstract training space. In addition, in every room of the house hangs the lamp Mayday (1999), which has been given, contrary to its otherwise variable and flexible use, a fixed bracket here, rounding off the installations and environments of each space like a safety utensil required by law.

Through these re-combinations of familiar objects, the show lets us experience intimations, sketches or prototype setups of "New Normals" – anticipations of one of many possible futures that might await us. In his design practice Konstantin Grcic dedicates himself to the shaping of the world we live in that is directed towards the future. Thus he constantly anticipates possible new states of the present in his pieces of furniture and other objects. In doing so, he actively shifts what is conceivable and translates it into the realm of what can be designed. Grcic's chairs, arm chairs, lamps, shelves and other objects provide impressive demonstrations of the fact that the future cannot be known – but it can be shaped. His designs shift our point of view from what already exists to what is coming.

In "New Normals" his designs not only leave their usually obvious function behind, but they also point to other worlds, new worlds, possible worlds, in which these objects can achieve their full potential. However, these experimental setups are not to be confused with prophecies. Instead, they alert the viewers to our present and the role the design of our everyday lives plays in it. What connection to our own lives, what expectations of the future – dystopian and utopian at the same time – can be read off these scenarios? What new approaches to and perspectives on the present do these new constellations entail?

Ludwig Engel and Anna Himmelsbach

KONSTANTIN GRCIC IN CONVERSATION
WITH LUDWIG ENGEL AND ANNA HIMMELSBACH

WHY IS THE EXHIBITION CALLED "NEW NORMALS"?

Konstantin Grcic: The title derives from the concept of "the new normal", which has entered common parlance during the pandemic. "The new normal" describes a form of future that has already become an everyday occurrence. It refers to circumstances or things that indicate our immediate reality having undergone some change. Wearing a mask on an ordinary day is no longer the exception but has become a new normality. "New Normals", then, does not relate directly to the ongoing pandemic, but poses a question beyond the present situation: what could be normal in the future and how would that affect the world of physical things we live with? The objects and installations on display are in essence speculations. The point is not to frame a vision of a specific future, but to outline and test existing fragments of possible future normalcies.

WILL OUR FUTURE ACTUALLY LOOK LIKE THE ONE YOU SKETCH WITH "NEW NORMALS"?

KG: From the outset it was important to me to understand the exhibition as a prospect of one among many possible futures. Design is always a projection, a forward motion. My initial research was informed by scenarios concerning climate change and its consequences as they are forecast based on scientific evidence. However, because this issue is extremely charged in a political sense, the show would have taken a turn I did not want it to take. "New Normals" is a design exhibition, after all. What is shown, and in which form, is under no obligation. The show's exhibits provide a myriad of starting points for one's own exploration of new normalcies. The partly re-purposed pieces of furniture, as well as all the objects and materials used in the installations, may seem strange at first glance. And yet they might suddenly be omnipresent and considered completely normal almost instantaneously.

THE POINTS OF DEPARTURE FOR EACH OF YOUR INSTALLATIONS ARE OBJECTS YOU HAVE DESIGNED. WHAT EFFECT DID THAT HAVE ON YOUR OWN RELATIONSHIP TO THOSE OBJECTS.

KG: Being an industrial designer, I develop products from the initial idea to mass production. Usually, this process takes between two and four years for each project. During this time, I form a very close relationship with these things - it might be said that the projects are like children I raise. As soon as the development phase is completed and the products are introduced to the market that personal bond to them dissolves. The things take on a life of their own and I turn to new projects. That is the way things work and it is okay. Sometimes I wish I was able to intervene once more, to play with the finished products and explore them further. This exhibition provided me with precisely that opportunity. I have changed my own products or reinterpreted them by staging them in a different form. Some of those products I have rebuilt from scratch, as it were, while others have been supplemented by extraneous elements. In the jargon of the computer age, this form of intervention might be called hacking.

DESIGN OFTEN ANTICIPATES THE FUTURE BY LENDING FORM TO WHAT IS COMING WHEN THAT DOES NOT HAVE A NAME OR A USE YET. IN MANY INSTANCES THE NEW EMERGES PRECISELY IN THOSE EVERYDAY MOMENTS WHEN REPURPOSING OR ALTERING AN EXISTING THING GENERATES SOMETHING THAT HAD NOT BEEN VISIBLE BEFORE. THAT IS ALSO HOW GOOD PIECES OF SCIENCE FICTION WORK...

KG: Our image of the future has long ceased to correspond to the hyper-design aesthetics of, say, Star Trek; Blade Runner would be a more

apt point of reference. It depicted the future as a real world, which is constantly under construction and is barely held together, and then mainly by duct tape. That picture of the future is not too unlike our own world. It is not redesigned all at once, but changes gradually through permanent mending and touching up as well as more substantial interventions. There is not one overarching design idea. Rather, there are many dynamics interacting with each other and melding into a heterogeneous whole.
The majority of what surrounds us today will remain the same for decades to come. Our cities and buildings will not look like spaceships, even Haus am Waldsee will, more or less, retain its current appearance. What differentiates the future from the present are many subtle details and small changes that, taken together, will shape our lives differently over the next five to ten or twenty years. While they are inconceivable today, they will be taken for granted then.

LESS FUTURE, MORE PRESENT?

KG: A future which pretends to be unambiguous is but a façade. The real future is incomplete and will always be a bit uncanny in a way. Observing everyday situations carefully provides me with the greatest inspiration in my work.
I am intrigued by the reality of our world as it is. It subjects us to conditions I, as designer, can challenge.

WHAT ROLE DOES THE DIGITAL WORLD PLAY FOR YOU WITHIN THIS CONTEXT?

KG: Everything is possible on the computer. Gravity does not apply and there are no obstacles to feasibility – just as there is no responsibility. You can put something big and heavy on tiny legs and it will still hold.
Pushing these freedoms to their limits digitally has become a crucial part of my design practice. Because industrial design is precisely about what is feasible, the computer gives me the opportunity to experiment, to think beyond the

certainties of the present moment. Working on "New Normals" was just such a field of experimentation in a physical form. We were able to implement ideas in reality that usually only exist on the screen. For me, the show will yield important new points of access and experiences, which I can apply in future projects.

YOU ARE REGULARLY CURATING AND DESIGNING EXHIBITIONS IN WHICH YOU ARE ABLE TO ACT WITH GREATER FREEDOM THAN IN A PROJECT FOR A CLIENT.

KG: Exhibition projects are not necessarily freer, but they are more complex than product design. I am talking about the various levels of design on which an exhibition operates: from the concept of the content, via the architecture and the presentation of the exhibits all the way to texts and media. As an exhibition designer I feel like a director, who focuses on the interplay of many different activities. An exhibition of one's own work raises an additional challenge: dealing with oneself. For "New Normals" I took existing works and transformed them into something new. The exhibits and installations on display allow me to shift my works even more into a direction where I would have liked to see them during the process of their development. And I am testing scenarios that did not take place during the stages of production.

SINCE WHEN HAVE EXHIBITIONS FORMED PART OF YOUR PRACTICE?

KG: I did the first exhibitions at the beginning of the 90s. The client was a company called "Authentics", a manufacturer of household goods made of plastic. They wanted commercial presentations on trade fairs. My stagings attempted to display the products not as goods for sale but in their cultural context and to tell their stories. In those days we worked with minimal budgets. The projects were to be realized in a simple and speedy fashion. Looking back, those were important, formative years for me. Designing exhibitions works like a well-crafted stage design. The design is reduced to just a few precise elements, succeeding largely because of what is not visible.

DID HAUS AM WALDSEE – A FORMER PRIVATE RESIDENCE WITH A SPECIFIC ARCHITECTURE – EXERT A PARTICULAR INFLUENCE ON THE SHOW?

KG: Without this particular building, "New Normals" would not exist in its current form. Haus am Waldsee was conceived as a luxurious private villa in the 1920s, a fact that is still very much in evidence. Since my work mainly consists of pieces of furniture it represents quite a challenge to create an exhibition at this specific site that does not turn out to be mere furnishing. I wanted to avoid that at all costs and nevertheless raise awareness that you find yourself in a formerly private residence. The house is the house and the objects are the objects. They interact with each other without making you think that the rooms were to be furnished or decorated.

WHAT WILL REMAIN OF "NEW NORMALS"?

KG: My hope is that the exhibition produces strong images, for those will remain. They are the images of individual exhibits, but also of their overall mis-en-scène at Haus am Waldsee. My wish is for the particular scenes to assemble into a larger atmospheric picture, which continues to have an effect beyond the space and duration of the exhibition. Perhaps we are able to encourage and maybe even trigger many of our visitors to examine their own images of the future and explore them further.

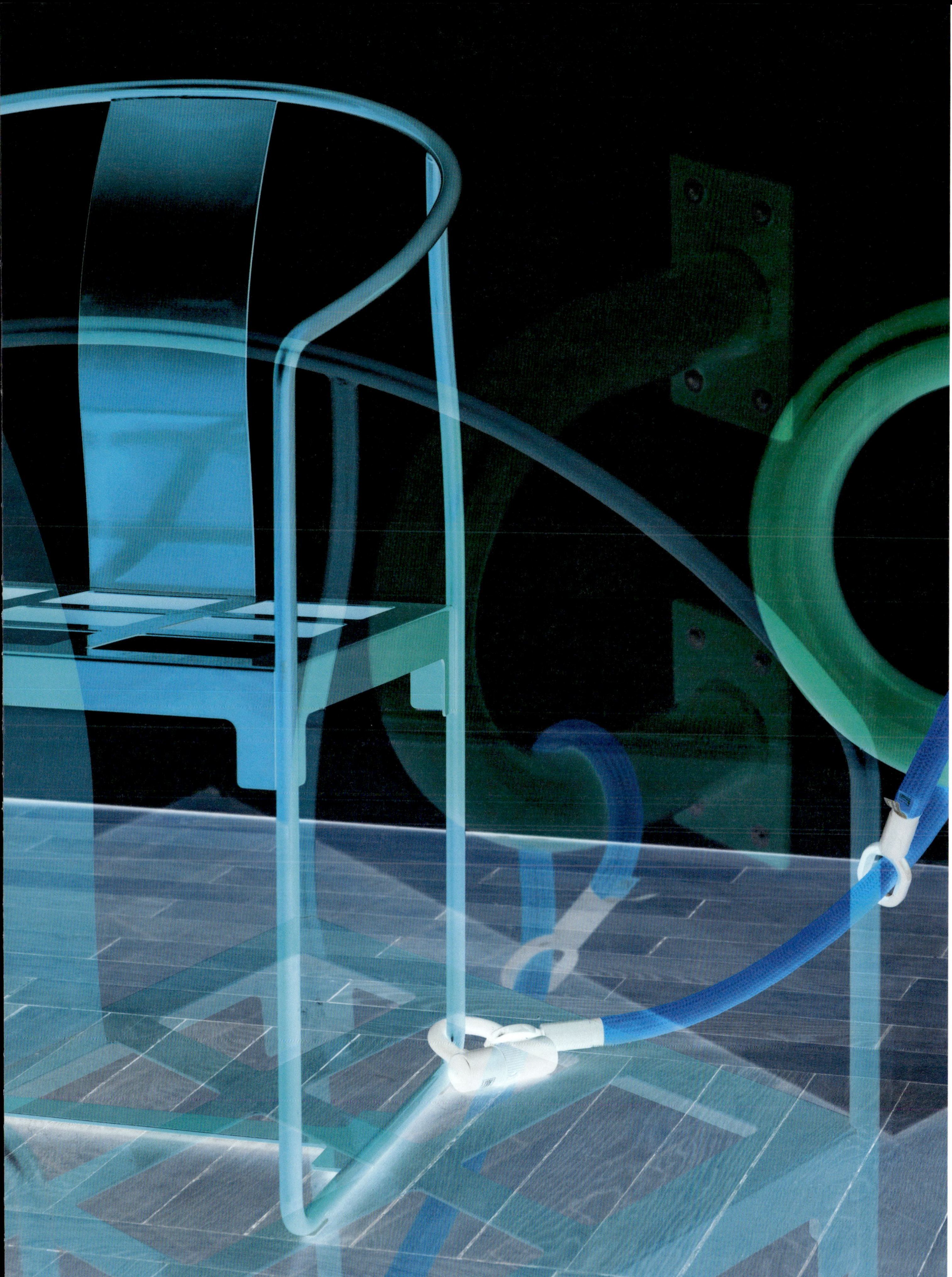

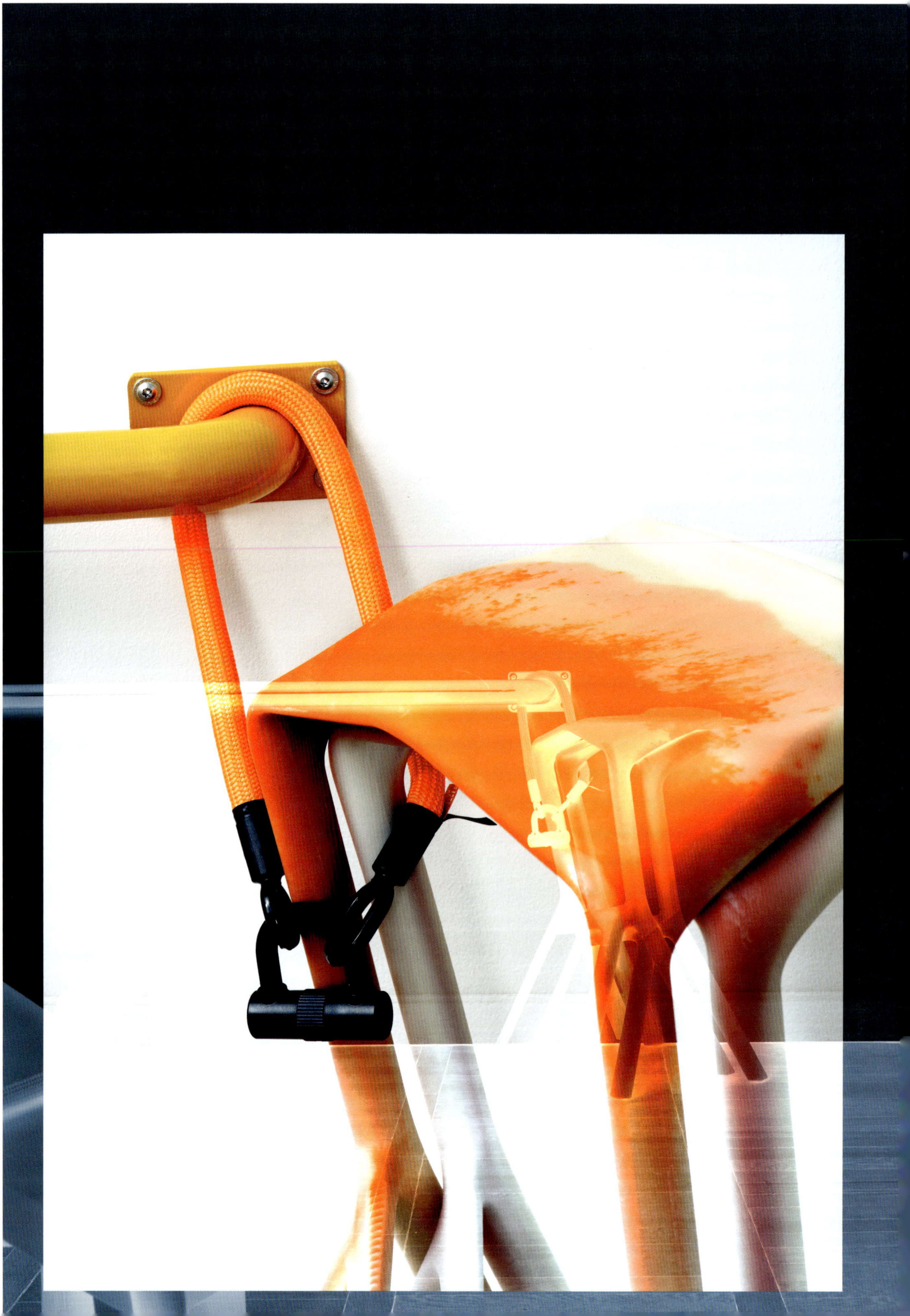

darkroom

KIZIO

14 24 48
25 DAY 1 MONTH 2028 YEAR TEMP 18 °C
SUN MON TUE WED THU FRI

CLASH
rage
METAL
VANS
"OFF THE WALL"
PALACE
SANTA CRUZ
Supreme
ASTRO BOY
MASTERMIND
JAPAN
THRASHER
RUN
DMC
OMP

FASTLANE

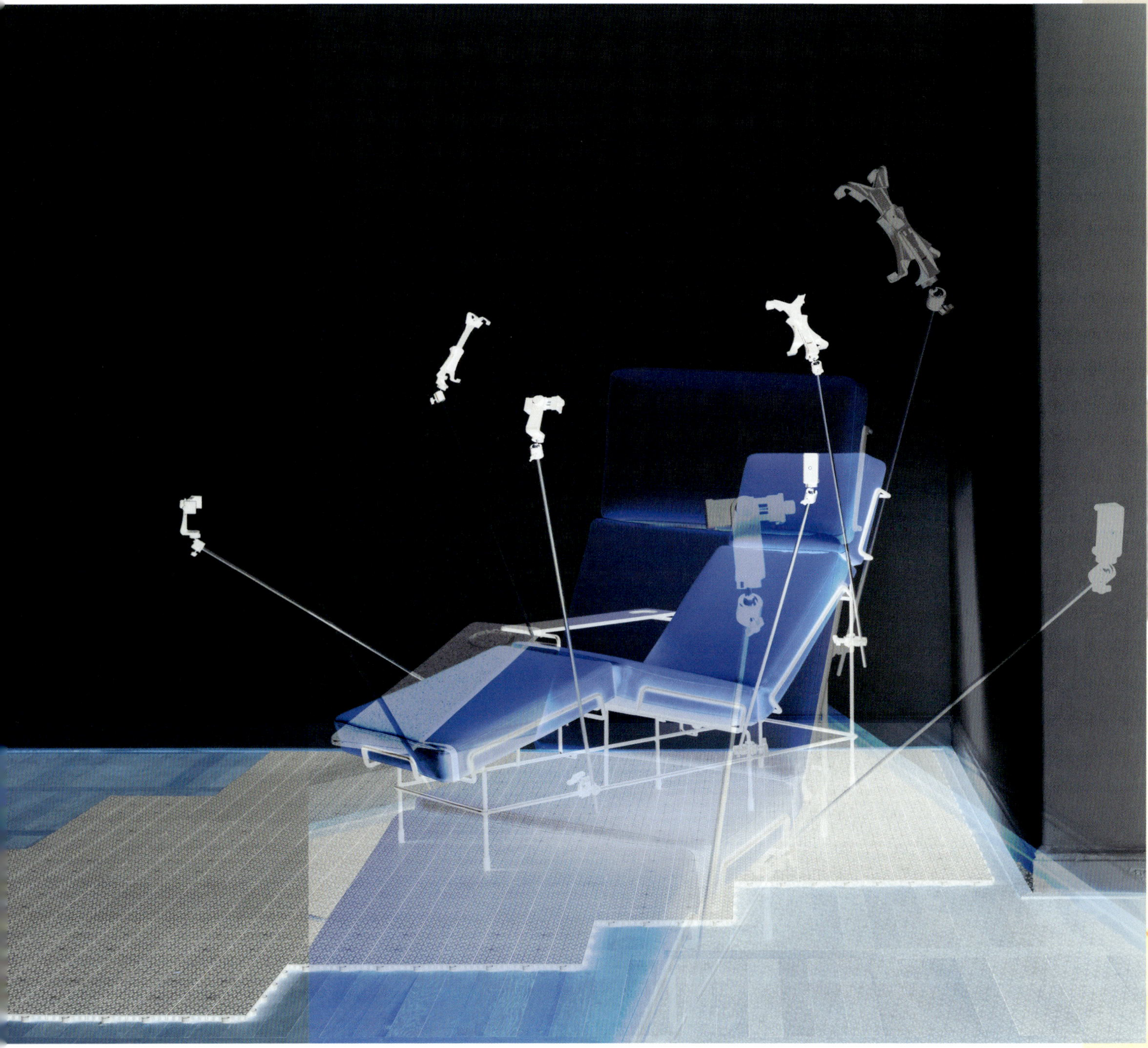

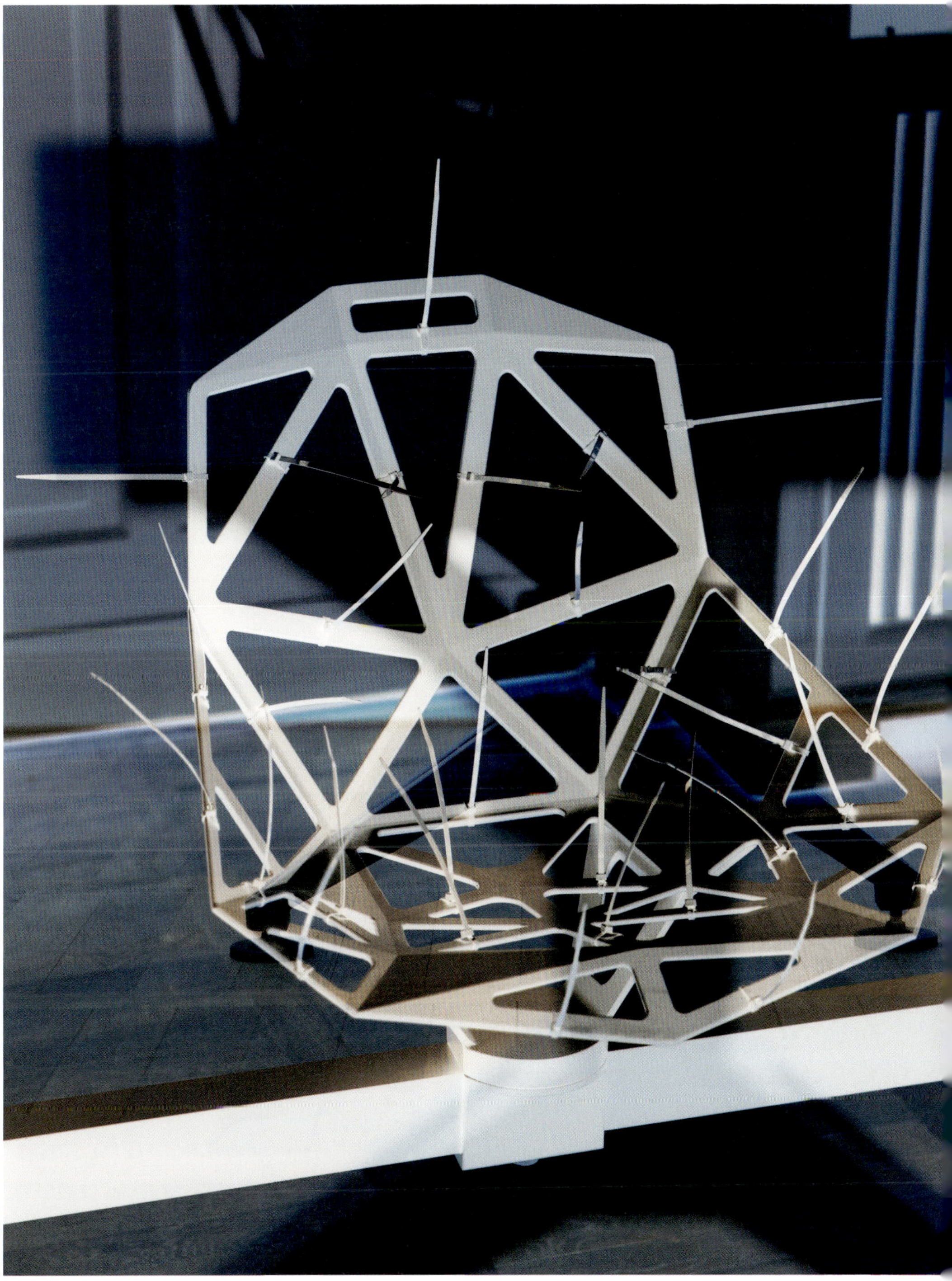

EliteAthlete

WERKLISTE

N■RMAL 0 S.15	MAYDAY Lampe / Flos / 1999 Blechgehäuse Steckdose
N■RMAL 1.1 S.16	BELL Monobloc Stuhl / Magis / 2020 Rammschutzbügel Fahrradschloss
N■RMAL 1.2 S.18, 20	MIURA Barhocker / Plank / 2005 Rammschutzbügel Fahrradschloss
N■RMAL 1.3 S.19	MINGX Stuhl / Driade / 2016 Rammschutzbügel Fahrradschloss
N■RMAL 2 S.22/23, 24, 25	MEDICI Lounge-Stuhl / Mattiazzi / 2012 Windschutzscheibe Citroën DS4 Rigipsplatten
N■RMAL 3 S.27, 28/29	ST■■L-T■■L Stuhl / Vitra / 2018 Radio- und TV-Zimmerantennen
N■RMAL 4 S.30, 31	SAM-S■N Sessel / Magis / 2015 Feuerwehrschlauch Bootsluke
N■RMAL 5 S.32/33, 34	BANZAI - CHAMPI■NS Tisch / Galerie kreo / 2011 Motorrad-Lenkerspiegel Neon-Glas
N■RMAL 6 S.36/37	HEAD IN THE SKY Prototyp / 2017 Satellitenschüsseln
N■RMAL 7 S.37	Kabelrinnen LED-Matrix
N■RMAL 8 S.38, 39	ALLSTAR Bürostuhl / Vitra / 2014 Schalungsstützen Gymnastikbänder Betonsteine
N■RMAL 9 S.41, 42	TABLE CHAIR Prototyp / Vitra / 2016 5-Punkte-Renngurt
N■RMAL 10 S.44, 45	360° CHAIR Stuhl / Magis / 2009 Aluminiumrohre Aluminium-Rohrverbinder Fahrzeug-Warnleuchte Bodenschwellen
N■RMAL 11 S.47, 48, 49	MIURA TABLE Tisch / Plank / 2007 MYT■ Stuhl / Plank 2008 Aluminium-Traversensystem Pantographen
N■RMAL 12 S.50, 51	TRAFFIC Chaiselongue / Magis / 2013 Smartphone- und Tablethalterungen Aluminiumstangen
N■RMAL 13 S.52, 53	PUBLIC_■NE Sitzbank / Magis / 2006 360° C■NTAINER Rollwagen / Magis / 2010 Kabelbinder
N■RMAL 14 S.55, 56/57	BRUT Sofa / Magis / 2016 Sicherheitsnetze Gymnastikball
N■RMAL 15 S.59, 60/61	CHAP Hocker / Vitra / 2021 Betonstein Kabelschlangen Blink-LED

LIST OF WORKS

NORMAL 0
p15
MAYDAY lamp / Flos / 1999
sheet metal casing
socket

NORMAL 1.1
p16
BELL Monobloc chair / Magis / 2020
crash protection barrier
bike lock

NORMAL 1.2
p18, 20
MIURA bar stool / Plank / 2005
crash protection barrier
bike lock

NORMAL 1.3
p19
MINGX chair / Driade / 2016
crash protection barrier
bike lock

NORMAL 2
p22/23, 24, 25
MEDICI lounge chair/ Mattiazzi / 2012
windscreen Citroën DS4
sheetrock

NORMAL 3
p27, 28/29
STOOL-TOOL chair / Vitra / 2018
radio and tv antennae

NORMAL 4
p30, 31
SAM-SON easy chair / Magis / 2015
fire hose
boat hatchway

NORMAL 5
p32/33, 34
BANZAI - CHAMPIONS table / Galerie kreo / 2011
bar end mirrors for motorbikes
neon glas

NORMAL 6
p36/37
HEAD IN THE SKY prototype / 2017
satellite dishes

NORMAL 7
p37
cable channels
LED matrix

NORMAL 8
p38, 39
ALLSTAR office chair/ Vitra / 2014
formwork props
gym band
concrete blocks

NORMAL 9
p41, 42
TABLE CHAIR prototype / Vitra / 2016
5-point racing harness

NORMAL 10
p44, 45
360° CHAIR chair / Magis / 2009
aluminium pipes
aluminium pipe connector
vehicle warning lights
ground beams

NORMAL 11
p47, 48, 49
MIURA TABLE table / Plank / 2007
MYTO chair / Plank 2008
aluminium cross-beam system
pantographs

NORMAL 12
p50, 51
TRAFFIC chaise longue / Magis / 2013
smartphone and tablet holders
aluminium sticks

NORMAL 13
p52, 53
PUBLIC_ONE bench / Magis / 2006
360° CONTAINER storage / Magis / 2010
cable tie

NORMAL 14
p55, 56/57
BRUT sofa / Magis / 2016
safety nets
exercise ball

NORMAL 15
p59, 60/61
CHAP stool / Vitra / 2021
concrete block
cable chain systems
flashing LED lights

BIOGRAFIE

Konstantin Grcic (*1965) erhielt eine Ausbildung zum Möbelbauer an der John Makepeace School for Craftsmen in Wood, ehe er Design am Royal College of Art in London studierte. Im Jahr 1991 eröffnete er sein eigenes Designbüro in München. Heute lebt und arbeitet Konstantin Grcic in Berlin. Grcic ist auf zahlreichen Gebieten aktiv, von Projekten zum Industriedesign über Ausstellungsdesign bis hin zu Kollaborationen mit Architekt*innen. Zu seinen berühmteren Kunden zählen Artek, Authentics, Cassina, ClassiCon, Flötotto, Flos, Galerie Kreo, Kettal, Laufen, Magis, Muji, Nespresso, Plank, Smart und Vitra. Er erhielt eine Reihe von Preisen, darunter der italienische Compasso d'Oro (2001, 2011, 2016) sowie Personality of the Year im Rahmen des deutschen Design Award (2016). Konstantin Grcic' Werk ist in mehreren Einzelausstellungen zu sehen gewesen, so im Haus der Kunst in München (2006); am Art Institute of Chicago (2009); im Vitra Design Museum in Weil am Rhein (2015) sowie in der Kunsthalle Bielefeld (2016). Seine Schöpfungen sind in den Sammlungen der wichtigsten Designmuseen der Welt zu finden: im Museum of Modern Art, New York; Victoria and Albert Museum, London; Centre Georges-Pompidou, Paris; Die Neue Sammlung, München. Grcic definiert Funktion mit Blick auf den Menschen und verbindet Strenge in der Form mit beachtlichem Scharfsinn und Humor. Charakteristisch für seine Arbeit ist sorgfältige Forschung hinsichtlich Kunst-, Design- und Architekturgeschichte sowie seine Leidenschaft für Technologie und Materialien.

BIOGRAPHY

Konstantin Grcic (*1965) was trained as a cabinet maker at The John Makepeace School for Craftsmen in Wood before studying Design at the Royal College of Art in London. He set up his own office in Munich in 1991. Today, Konstantin Grcic is based in Berlin. Grcic is active in several fields ranging from industrial design projects, exhibition design and architectural collaborations. Amongst his renowned clients are Artek, Authentics, Cassina, ClassiCon, Flötotto, Flos, Galerie Kreo, Kettal, Laufen, Magis, Muji, Nespresso, Plank, Smart and Vitra. He is the recipient of numerous awards such as the Italian Compasso d'Oro (2001, 2011, 2016) and the German Design Award Personality Award (2016). Konstantin Grcic's work has been showcased in various solo exhibitions, including Haus der Kunst, Munich (2006); Art Institute of Chicago (2009); Vitra Design Museum, Weil am Rhein (2015) and Kunsthalle Bielefeld (2016). His creations form part of the permanent collections of the world's most important design museums such as the Museum of Modern Art, New York; Victoria and Albert Museum, London; Centre Georges-Pompidou, Paris; Die Neue Sammlung, Munich. Grcic defines function in human terms, combining formal strictness with considerable mental acuity and humour. His work is characterised by a careful research into the history of art, design and architecture and his passion for technology and materials.

DANK

Von Herzen danken wir Konstantin Grcic für seinen unermüdlichen Einsatz und das Engagement, das er in vielen Treffen und Gesprächen mit uns auf der Suche nach der richtigen Form für seine New Normals an den Tag gelegt hat.
Wir danken dem Vorstand des Trägervereins, Jakob Braeuer, Kaspar von Erffa und Leonie von Gadow, für das Vertrauen in uns. Wir danken außerdem dem Team von Konstantin Grcic, allen voran Kathrin Hasskamp und Tobias Trübenbacher. Ein ganz besonderer Dank geht an Florian Böhm und Studio AKFB. Der Fotograf begleitet Konstantin Grcic schon seit Beginn seines Schaffens, und so waren wir auch bei der Katalogentwicklung schnell ein eingespieltes Team und hatten viel Freude an der Zusammenarbeit. Zudem bedanken wir uns bei dem Team des Hauses am Waldsee, Tobias Bader, dem Geschäftsführer, Susanne Witt, der Assistentin der Geschäftsführung, Erik Günther für die Pressearbeit, Laura Kamlade für die digitale Kommunikation, Katherina Perlongo für die Kunstvermittlung, Uria Man für die Organisation der Veranstaltungen, Isabel Grahn, unserer studentischen Hilfskraft, und Nele Schinz für die engagierte Pflege des Skulpturengartens. Nicht zuletzt danken wir unserem Kassen- und Aufsichtsteam, Jonas Bergmann, Michael Gottberg, Annette Kieselbach und Roswitha Wielgosch. Unser Dank geht außerdem an Selma Sanner und Maximilian Riedel von der Agentur Boros. Für das Lektorat danken wir Şebnem Yavuz und Volker Ellerbeck für die Übersetzung.
Für die Finanzierung der Ausstellung sei dem Hauptstadtkulturfonds und dem Verein der Freunde und Förderer des Hauses am Waldsee gedankt. Für die Förderung des Hauses danken wir der Senatsverwaltung für Kultur und Europa sowie dem Bezirk Steglitz-Zehlendorf von Berlin. Nicht zuletzt bedanken wir uns bei den Firmen ClassiCon, Dirade, Flos, Magis, Plank und Vitra, die uns Objekte von Konstantin Grcic so freundlich zur Verfügung gestellt haben.

THANKS

We are deeply grateful to Konstantin Grcic for the tireless efforts and commitment he has shown in numerous meetings and conversations with us in search of the right form of his New Normals.
We thank the board of the supporting association, Jakob Braeuer, Kaspar von Erffa and Leonie von Gadow, for the trust they place in us. Moreover, we thank Konstantin Grcic' team, with Kathrin Hasskamp and Tobias Trübenbacher leading the way. Special thanks go to Florian Böhm and Studio AKFB. The photographer has accompanied Konstantin Grcic from his beginnings, which quickly lead to the well-coordinated development of the catalogue, a collaboration that gave everyone concerned enormous joy. Further thanks go to the team at Haus am Waldsee, Tobias Bader, managing director; Susanne Witt, assistant managing director; Erik Günther, press and communications; Laura Kamlade, digital communication; Katherina Perlongo, art education; Uria Man, event manager; Isabel Grahn, student associate; and Nele Schinz for taking great care of the sculpture garden. Not least we are indebted to our team of front desk officers and invigilators, Jonas Bergmann, Michael Gottberg, Annette Kieselbach and Roswitha Wielgosch. In addition, thanks are due to Selma Sanner and Maximilian Riedel of the agency Boros. Thanks also to Şebnem Yavuz for copy-editing and Volker Ellerbeck for translating the texts in this catalogue.
For financial support of the exhibition we are indebted to German Capital Cultural Fund (HKF) as well as the Association of Friends and Supporters of Haus am Waldsee. The day-to-day business of Haus am Waldsee is supported by Senate Department for Culture and Europe and the borough of Berlin Steglitz-Zehlendorf; we are grateful to both institutions. Finally, we thank the companies ClassiCon, Dirade, Flos, Magis, Plank and Vitra for their kind and generous loans of objects by Konstantin Grcic for the purposes of this exhibition.

IMPRESSUM | IMPRINT

Diese Publikation erscheint anlässlich der Ausstellung | This catalogue is published on the occasion of the exhibition

Konstantin Grcic – New Normals
21. Januar 2022 – 8. Mai 2022
January 21, 2022 – May 8, 2022

KATALOG | PUBLICATION

Für das | For the Haus am Waldsee herausgegeben von | edited by:
Ludwig Engel, Anna Himmelsbach
Texte | Texts:
Ludwig Engel, Anna Himmelsbach, Konstantin Grcic
Lektorat | Copy-editing:
Şebnem Yavuz
Übersetzung | Translation:
Volker Ellerbeck
Gestaltung | Design:
Studio AKFB
Fotograf | Photographer:
Florian Böhm
Schrift | Font:
KGHAW by Studio AKFB
Druck | Print:
Druckerei Vogl, München

Gedruckt | Printed in Deutschland | Germany
Auflage | edition of: 800

Vertrieb | Distribution:
Buchhandlung Walther König
Ehrenstr. 4
D-50672 Köln
T +49 (0) 221 / 20 59 6 53
verlag@buchhandlung-walther-koenig.de

ISBN 978-3-7533-0187-7

Umschlagabbildung | Cover image:
Foto | Photo:
Florian Böhm

AUSSTELLUNG | EXHIBITION

Kuratiert von | Curated by:
Ludwig Engel, Anna Himmelsbach, Konstantin Grcic
Geschäftsführung | Managing Director:
Tobias Bader
Assistenz der Geschäftsführung | Assistant to Managing Director:
Susanne Witt
Projektmanagement | Project Management:
Anna Himmelsbach, Team KGID: Kathrin Hasskamp, Tobias Trübenbacher
Presse und Kommunikation | Press and Communications:
Erik Günther
Digitale Kommunikation | Digital Communications:
Laura Kamlade
Kunstvermittlung | Art Education:
Katherina Perlongo
Studentische Hilfskraft | Student assistant:
Isabel Grahn
Aufbau | Installation:
Lukas Bugla, Sylvan Dahlgrün, Ronny Lischinski; Team KGID: Simon Frambach, Pascal Hien, Frederic Rätsch, Tobias Trübenbacher
Kasse und Aufsicht | Front Desk and Invigilation:
Jonas Bergmann, Michael Gottberg, Annette Kieselbach, Roswitha Wielgosch
Trailer | Trailer:
Ole Schwarz
Ausstellungsgestaltung | Exhibition design:
KGID
Ausstellungsgrafik | Exhibition graphic:
Studio AKFB und | and KGID

Haus am Waldsee
Internationale Kunst in Berlin
Argentinische Allee 30
D-14163 Berlin
T +49 (0) 30 / 80 18 935
info@hausamwaldsee.de
www.hausamwaldsee.de

DAS HAUS AM WALDSEE WIRD GETRAGEN DURCH | HAUS AM WALDSEE IS LED BY
Haus am Waldsee e.V.
Vorstand | Board of trustees:
Jakob Braeuer
Kaspar von Erffa
Leonie von Gadow

DAS HAUS AM WALDSEE WIRD GEFÖRDERT DURCH | HAUS AM WALDSEE IS SUPPORTED BY

Senatsverwaltung für Kultur und Europa BERLIN
Fachbereich Kultur Steglitz Zehlendorf

DIE AUSSTELLUNG WIRD GEFÖRDERT DURCH | THE EXHIBTION IS SUPPORTED BY

HAUPT STADT KULTUR FONDS
HAUS AM WALDSEE FREUNDE UND FÖRDERER

WIR BEDANKEN UNS BEI UNSEREN PARTNERN UND SPONSOREN | WE WOULD LIKE TO THANK OUR PARTNERS AND SPONSORS

EIDOTECH www.eidotech.com
Kleemann & Wollmann Partnerschaftsgesellschaft mbB Steuerberater • Wirtschaftsprüfer • Rechtsanwalt
PARFÜMERIE Harbeck SEIT 1938
HOFER ANTIKSCHMUCK
BUCHHANDLUNG BRAUN & HASSENPFLUG

MEDIENPARTNER | MEDIA PARTNERS

rbb KULTUR
tipBerlin